THE ORIGIN OF HUMAN RACES
AND THE ANTIQUITY OF MAN DEDUCED
FROM THE THEORY OF
"NATURAL SELECTION"
(1864)

BY

ALFRED RUSSEL WALLACE

British Library Cataloguing-in-Publication Data
A catalogue record for this book is available from the
British Library

Alfred Russel Wallace

Alfred Russel Wallace was born on 8th January 1823 in the village of Llanbadoc, in Monmouthshire, Wales.

At the age of five, Wallace's family moved to Hertford where he later enrolled at Hertford Grammar School. He was educated there until financial difficulties forced his family to withdraw him in 1836. He then boarded with his older brother John before becoming an apprentice to his eldest brother, William, a surveyor. He worked for William for six years until the business declined due to difficult economic conditions.

After a brief period of unemployment, he was hired as a master at the Collegiate School in Leicester to teach drawing, map-making, and surveying. During this time he met the entomologist Henry Bates who inspired Wallace to begin collecting insects. He and bates continued exchanging letters after Wallace left teaching to pursue his surveying career. They corresponded on prominent works of the time such as Charles Darwin's *The Voyage of the Beagle* (1839) and Robert Chamber's *Vestiges of the Natural History of Creation* (1844).

Wallace was inspired by the travelling naturalists of the day and decided to begin his exploration career collecting specimens in the Amazon rainforest. He explored the Rio Negra for four years, making notes on the peoples and languages he encountered as well as the geography, flora, and fauna. On his return voyage his ship, Helen, caught fire and he and the crew were stranded for ten days before being picked up by the Jordeson, a brig travelling from Cuba to London. All of his specimens aboard Helen had been lost.

After a brief stay in England he embarked on a journey to the Malay Archipelago (now Singapore, Malaysia, and Indonesia). During this eight year period he collected more than 126,000 specimens, several thousand of which represented new species to science. While travelling, Wallace refined his thoughts about evolution and in 1858 he outlined his theory of natural selection in an article he sent to Charles Darwin. This was published in the same year along with Darwin's own theory. Wallace eventually published an account of his travels *The Malay Archipelago* in 1869, and it became one of the most popular books of scientific exploration in the 19th century.

Upon his return to England, in 1862, Wallace became a staunch defender of Darwin's landmark work *On the Origin of Species* (1859). He wrote responses to those critical of the theory of natural selection, including 'Remarks on the Rev. S. Haughton's Paper on the Bee's Cell, And on the Origin of Species' (1863) and 'Creation by Law' (1867). The former of these was particularly pleasing to Darwin. Wallace also published important papers such as 'The Origin of Human Races and the Antiquity of Man Deduced from the Theory of 'Natural Selection" (1864) and books, including the much cited *Darwinism* (1889).

Wallace made a huge contribution to the natural sciences and he will continue to be remembered as one of the key figures in the development of evolutionary theory.

Wallace died on 7th November 1913 at the age of 90. He is buried in a small cemetery at Broadstone, Dorset, England.

THE ORIGIN OF HUMAN RACES AND THE ANTIQUITY OF MAN DEDUCED FROM THE THEORY OF "NATURAL SELECTION"

Among the most advanced students of man, there exists a wide difference of opinion on some of the most vital questions respecting his nature and origin. Anthropologists are now, indeed, pretty well agreed that man is not a recent introduction into the earth. All who have studied the question now admit that his antiquity is very great; and that, though we have to some extent ascertained the minimum of time during which he *must* have existed, we have made no approximation towards determining that far greater period during which he *may* have, and probably *has*, existed. We can with tolerable certainty affirm that man must have inhabited the earth a thousand centuries ago, but we cannot assert that he positively did not exist, or that there is any good evidence against his having existed, for a period of a hundred thousand centuries. We know positively that he was contemporaneous with many now extinct animals, and has survived changes of the earth's surface fifty or a hundred times greater than any that have occurred during the historical period; but we cannot place any definite limit to the number of species he may have outlived, or to the amount of terrestrial change he may have witnessed.

But while on this question of man's antiquity there is a very general agreement,--and all are waiting eagerly for fresh evidence to clear up those points which all admit to be full of doubt,--on other and not less obscure and difficult questions a considerable amount of dogmatism is exhibited; doctrines are put forward as established truth, no doubt or hesitation is admitted, and it seems to be supposed that no further evidence is required, or that any new facts can modify our convictions. This is especially the case when we inquire, *Are the various forms under which man now exists primitive, or derived from preexisting forms; in other words, is man of one or many species?* To this question we immediately obtain distinct answers diametrically opposed to each other: the one party positively maintaining that man is a *species* and is essentially *one*--that all differences are but local and temporary variations, produced by the different physical and moral conditions by which he is surrounded; the other party maintaining with equal confidence that man is a genus of *many species*, each of which is practically unchangeable, and has ever been as distinct, or even more distinct, than we now behold them. This difference of opinion is somewhat remarkable, when we consider that both parties are well acquainted with the subject; both use the same vast accumulation of facts; both reject those early traditions of mankind which profess to give an account of his origin; and both declare that they are seeking fearlessly after truth alone. I believe, however, it will be found to be the old story over again of the shield--gold on one side and silver on the other--about which the knights disputed; each party will persist in looking only at the portion of truth on his own side of the question, and at the error which is mingled with his opponent's doctrine. It is my wish to show how the two opposing views can be combined so as to eliminate the error and retain the truth in each, and it is by means of Mr. Darwin's celebrated theory of "Natural Selection" that I hope to do this, and thus to harmonise the conflicting theories of modern anthropologists.

Let us first see what each party has to say for itself. In favour of the unity of mankind it is argued that there are no races without transitions to others; that every race exhibits within itself variations of colour, of hair, of feature, and of form, to such a degree as to bridge over to a large extent the gap that separates it from other races. It is asserted that no race is homogeneous; that there is a tendency to vary; that climate, food, and habits produce and render permanent physical peculiarities, which, though slight in the limited periods allowed to our observation, would, in the long ages during which the human race has existed, have sufficed to produce all the differences that now appear. It is further asserted that the advocates of the opposite theory do not agree among themselves; that some would make three, some five, some fifty or a hundred and fifty species of man; some would have had each species created in pairs, while others require nations to have at once sprung into existence, and that there is no stability or consistency in any doctrine but that of one primitive stock.

The advocates of the original diversity of man, on the other hand, have much to say for themselves. They argue that proofs of change in man have never been brought forward except to the most trifling amount, while evidence of his permanence meets us everywhere. The Portuguese and Spaniards, settled for two or three centuries in South America, retain their chief physical, mental, and moral characteristics; the Dutch boers at the Cape, and the descendants of the early Dutch settlers in the Moluccas, have not lost the features or the colour of the Germanic races; the Jews, scattered over the world in the most diverse climates, retain the same characteristic lineaments everywhere; the Egyptian sculptures and paintings show us that, for at least 4000 or 5000 years, the strongly contrasted features of the Negro and the Semitic races have remained altogether unchanged; while more recent discoveries prove that, in the case at least of the American aborigines, the mound-builders of the Mississippi valley, and the dwellers on Brazilian mountains, had still in the very infancy of the human race the same characteristic type of cranial formation that now distinguishes them.

If we endeavour to decide impartially on the merits of

this difficult controversy, judging solely by the evidence that each party has brought forward, it certainly seems that the best of the argument is on the side of those who maintain the primitive diversity of man. Their opponents have not been able to refute the permanence of existing races as far back as we can trace them, and have failed to show, in a single case, that at any former epoch the well marked varieties of mankind approximated more closely than they do at the present day. At the same time this is but negative evidence. A condition of immobility for four or five thousand years, does not preclude an advance at an earlier epoch, and--if we can show that there are causes in nature which would check any further physical change when certain conditions were fulfilled--does not even render such an advance improbable, if there are any general arguments to be adduced in its favour. Such a cause, I believe, does exist, and I shall now endeavour to point out its nature and its mode of operation.

In order to make my argument intelligible, it is necessary for me to explain very briefly the theory of "Natural Selection" promulgated by Mr. Darwin, and the power which it possesses of modifying the forms of animals and plants. The grand feature in the multiplication of organic life is that of close general resemblance, combined with more or less individual variation. The child resembles its parents or ancestors more or less closely in all its peculiarities, deformities, or beauties; it resembles them in general more than it does any other individuals; yet children of the same parents are not all alike, and it often happens that they differ very considerably from their parents and from each other. This is equally true of man, of all animals, and of all plants. Moreover, it is found that individuals do not differ from their parents in certain particulars only, while in all others they are exact duplicates of them. They differ from them and from each other in every particular: in form, in size, in colour, in the structure of internal as well as of external organs; in those subtle peculiarities which produce differences of constitution, as well as in those still more subtle ones which lead to modifications of mind and character. In other words, in every possible way, in every organ and in every function, individuals of the same stock vary.

Now, health, strength, and long life are the results of a harmony between the individual and the universe that surrounds it. Let us suppose that at any given moment this harmony is perfect. A certain animal is exactly fitted to secure its prey, to escape from its enemies, to resist the inclemencies of the seasons, and to rear a numerous and healthy offspring. But a change now takes place. A series of cold winters, for instance, come on, making food scarce, and bringing an immigration of some other animals to compete with the former inhabitants of the district. The new immigrant is swift of foot, and surpasses its rivals in the pursuit of game; the winter nights are colder, and require a thicker fur as a protection, and more nourishing food to keep up the heat of the system. Our supposed perfect animal is no longer in harmony with its universe; it is in danger of dying of cold or of starvation. But the animal varies in its offspring. Some of these are swifter than others--they still manage to catch food enough; some are hardier and more thickly furred-- they manage in the cold nights to keep warm enough; the slow, the weak, and the thinly clad soon die off. Again and again, in each succeeding generation, the same thing takes place. By this natural process, which is so inevitable that it cannot be conceived not to act, those best adapted to live, live; those least adapted, die. It is sometimes said that we have no direct evidence of the action of this selecting power in nature. But it seems to me we have better evidence than even direct observation would be, because it is more universal, viz., the evidence of necessity. It must be so; for, as all wild animals increase in a geometrical ratio, while their actual numbers remain on the average stationary, it follows that as many die annually as are born. If therefore, we deny natural selection, it can only be by asserting that in such a case as I have supposed, the strong, the healthy, the swift, the well clad, the well organised animals in every respect, have no advantage over,--do not on the average live longer than the weak, the unhealthy, the slow, the ill-clad, and the imperfectly organised individuals; and this no sane man has yet been found hardy enough to assert. But this is not all; for the offspring on the average resemble their parents, and the selected portion of each succeeding generation will therefore be stronger, swifter, and more thickly furred than the last; and if this process goes on for thousands of generations, our animal will have again become thoroughly in harmony with the new conditions in which he is placed. But he will now be a different creature. He will be not only swifter and stronger, and more furry, he will also probably have changed in colour, in form, perhaps have acquired a longer tail, or differently shaped ears; for it is an ascertained fact, that when one part of an animal is modified, some other parts almost always change as it were in sympathy with it. Mr. Darwin calls this *"correlation of growth,"* and gives as instances that hairless dogs have imperfect teeth; blue eyed cats are deaf; small feet accompany short beaks in pigeons; and other equally interesting cases.

Grant, therefore, the premises: 1st. That peculiarities of every kind are more or less hereditary. 2nd. That the offspring of every animal vary more or less in all parts of their organisation. 3rd. That the universe in which these animals live, is not absolutely invariable;--none of which propositions can be denied; and then consider that the animals in any country (those at least which are not dying out) must at each successive period be brought into harmony with the surrounding conditions; and we have all the elements for a change of form and structure in the animals, keeping exact pace with changes of whatever nature in the surrounding universe. Such changes must be slow, for the changes in the universe are very slow; but just as these slow changes become important, when we look at results after long periods of action, as we do when we perceive the alterations of the earth's surface during geological epochs; so the parallel changes in animal form become more and more striking according as the time they have been going on is great, as we see when we compare our living animals with those which we disentomb from each successively older geological formation.

This is briefly the theory of "natural selection," which explains the changes in the organic world as being parallel with, and in part dependent on those in the inorganic. What we now have to inquire is,--Can this theory be applied in any way to the question of the origin of the races of man? or is there anything in human nature that takes him out of the category of those organic existences, over whose successive mutations it has had such powerful sway?

In order to answer these questions, we must consider why it is that "natural selection" acts so powerfully upon animals, and we shall, I believe, find that its effect depends mainly upon their self-dependence and individual isolation. A slight injury, a temporary illness, will often end in death, because it leaves the individual powerless against its enemies. If a herbivorous animal is a little sick and has not fed well for a day or two, and the herd is then pursued by a beast of prey, our poor invalid inevitably falls a victim. So in a carnivorous animal the least deficiency of vigour prevents its capturing food, and it soon dies of starvation. There is, as a general rule, no mutual assistance between adults, which enables them to tide over a period of sickness. Neither is there any division of labour; each must fulfil *all* the conditions of its existence, and, therefore, "natural selection" keeps all up to a pretty uniform standard.

But in man, as we now behold him, this is different. He is social and sympathetic. In the rudest tribes the sick are assisted at least with food; less robust health and vigour than the average does not entail death. Neither does the want of perfect limbs or other organs produce the same effects as among animals. Some division of labour takes place; the swiftest hunt, the less active fish, or gather fruits; food is to some extent exchanged or divided. The action of natural selection is therefore checked; the weaker, the dwarfish, those of less active limbs, or less piercing eyesight, do not suffer the extreme penalty which falls upon animals so defective.

In proportion as these physical characteristics become of less importance, mental and moral qualities will have increasing influence on the well-being of the race. Capacity for acting in concert, for protection and for the acquisition of food and shelter; sympathy, which leads all in turn to assist each other; the sense of right, which checks depredations upon our fellows; the decrease of the combative and destructive propensities; self-restraint in present appetites; and that intelligent foresight which prepares for the future, are all qualities that from their earliest appearance must have been for the benefit of each community, and would, therefore, have become the subjects of "natural selection." For it is evident that such qualities would be for the well-being of man; would guard him against external enemies, against internal dissensions, and against the effects of inclement seasons and impending famine, more surely than could any merely physical modification. Tribes in which such mental and moral qualities were predominant, would therefore have an advantage in the struggle for existence over other tribes in which they were less developed, would live and maintain their numbers, while the others would decrease and finally succumb.

Again, when any slow changes of physical geography, or of climate, make it necessary for an animal to alter its food, its clothing, or its weapons, it can only do so by a corresponding change in its own bodily structure and internal organisation. If a larger or more powerful beast is to be captured and devoured, as when a carnivorous animal which has hitherto preyed on sheep is obliged from their decreasing numbers to attack buffaloes, it is only the strongest who can hold,--those with most powerful claws, and formidable canine teeth, that can struggle with and overcome such an animal. Natural selection immediately comes into play, and by its action these organs gradually become adapted to their new requirements. But man, under similar circumstances, does not require longer nails or teeth, greater bodily strength or swiftness. He makes sharper spears, or a better bow, or he constructs a cunning pitfall, or combines in a hunting party to circumvent his new prey. The capacities which enable him to do this are what he requires to be strengthened, and these will, therefore, be gradually modified by "natural selection," while the form and structure of his body will remain unchanged. So when a glacial epoch comes on, some animals must acquire warmer fur, or a covering of fat, or else die of cold. Those best clothed by nature are, therefore, preserved by natural selection. Man, under the same circumstances, will make himself warmer clothing, and build better houses; and the necessity of doing this will react upon his mental organisation and social condition--will advance them while his natural body remains naked as before.

When the accustomed food of some animal becomes scarce or totally fails, it can only exist by becoming adapted to a new kind of food, a food perhaps less nourishing and less digestible. "Natural selection" will now act upon the stomach and intestines, and all their individual variations will be taken advantage of to modify the race into harmony with its new food. In many cases, however, it is probable that this cannot be done. The internal organs may not vary quick enough, and then the animal will decrease in numbers, and finally become extinct. But man guards himself from such accidents by superintending and guiding the operations of nature. He plants the seed of his most agreeable food, and thus procures a supply independent of the accidents of varying seasons or natural extinction. He domesticates animals which serve him either to capture food or for food itself, and thus changes of any great extent in his teeth or digestive organs are rendered unnecessary. Man, too, has everywhere the use of fire, and by its means can render palatable a variety of animal and vegetable substances, which he could hardly otherwise make use of, and thus obtains for himself a supply of food far more varied and abundant than that which any animal can command.

Thus man, by the mere capacity of clothing himself, and making weapons and tools, has taken away from nature that power of changing the external form and structure which she exercises over all other animals. As the competing races by which they are surrounded, the climate, the vegetation, or the animals which serve them for food, are slowly changing, they must undergo a corresponding change in their structure, habits, and constitution, to keep them in harmony with the new conditions--to enable them to live and maintain their numbers. But man does this by means of his intellect alone; which enables him with an unchanged body still to keep in harmony with the changing universe.

From the time, therefore, when the social and sympathetic

feelings came into active operation, and the intellectual and moral faculties became fairly developed, man would cease to be influenced by "natural selection" in his physical form and structure; as an animal he would remain almost stationary; the changes of the surrounding universe would cease to have upon him that powerful modifying effect which it exercises over other parts of the organic world. But from the moment that his body became stationary, his mind would become subject to those very influences from which his body had escaped; every slight variation in his mental and moral nature which should enable him better to guard against adverse circumstances, and combine for mutual comfort and protection, would be preserved and accumulated; the better and higher specimens of our race would therefore increase and spread, the lower and more brutal would give way and successively die out, and that rapid advancement of mental organisation would occur, which has raised the very lowest races of man so far above the brutes (although differing so little from some of them in physical structure), and, in conjunction with scarcely perceptible modifications of form, has developed the wonderful intellect of the Germanic races.

But from the time when this mental and moral advance commenced, and man's physical character became fixed and immutable, a new series of causes would come into action, and take part in his mental growth. The diverse aspects of nature would now make themselves felt, and profoundly influence the character of the primitive man.[1]

When the power that had hitherto modified the body, transferred its action to the mind, then races would advance and become improved merely by the harsh discipline of a sterile soil and inclement seasons. Under their influence, a hardier, a more provident, and a more social race would be developed, than in those regions where the earth produces a perennial supply of vegetable food, and where neither foresight nor ingenuity are required to prepare for the rigours of winter. And is it not the fact that in all ages, and in every quarter of the globe, the inhabitants of temperate have been superior to those of tropical countries? All the great invasions and displacements of races have been from North to South, rather than the reverse; and we have no record of there ever having existed, any more than there exists to-day, a solitary instance of an indigenous inter-tropical civilisation. The Mexican civilisation and government came from the North, and, as well as the Peruvian, was established, not in the rich tropical plains, but on the lofty and sterile plateaux of the Andes. The religion and civilisation of Ceylon were introduced from North India; the successive conquerors of the Indian peninsula came from the North-west, and it was the bold and adventurous tribes of the North that overran and infused new life into Southern Europe.

It is the same great law of "*the preservation of favoured races in the struggle for life,"* which leads to the inevitable extinction of all those low and mentally undeveloped populations with which Europeans come in contact. The red Indian in North America, and in Brazil; the Tasmanian, Australian and New Zealander in the southern hemisphere, die out, not from any one special cause, but from the inevitable effects of an unequal mental and physical struggle. The intellectual and moral, as well as the physical qualities of the European are superior; the same powers and capacities which have made him rise in a few centuries from the condition of the wandering savage[2] with a scanty and stationary population to his present state of culture and advancement, with a greater average longevity, a greater average strength, and a capacity of more rapid increase,--enable him when in contact with the savage man, to conquer in the struggle for existence, and to increase at his expense, just as the more favourable increase at the expense of the less favourable varieties in the animal and vegetable kingdoms, just as the weeds of Europe overrun North America and Australia, extinguishing native productions by the inherent vigour of their organisation, and by their greater capacity for existence and multiplication.

If these views are correct; if in proportion as man's social, moral and intellectual faculties became developed, his physical structure would cease to be affected by the operation of "natural selection," we have a most important clue to the origin of races. For it will follow, that those striking and constant peculiarities which mark the great divisions of mankind, could not have been produced and rendered permanent after the action of this power had become transferred from physical to mental variations. They must, therefore, have existed since the very infancy of the race; they must have originated at a period when man was gregarious, but scarcely social, with a mind perceptive but not reflective, ere any sense of *right* or feelings of *sympathy* had been developed in him.

By a powerful effort of the imagination, it is just possible to perceive him at that early epoch existing as a single homogeneous race without the faculty of speech, and probably inhabiting some tropical region. He would be still subject, like the rest of the organic world, to the action of "natural selection," which would retain his physical form and constitution in harmony with the surrounding universe. He must have been even then a dominant race, spreading widely over the warmer regions of the earth as it then existed, and, in agreement with what we see in the case of other dominant species, gradually becoming modified in accordance with local conditions. As he ranged farther from his original home, and became exposed to greater extremes of climate, to greater changes of food, and had to contend with new enemies, organic and inorganic, useful variations in his constitution would be selected and rendered permanent, and would, on the principle of "correlation of growth", be accompanied [[p. clxvi]] by corresponding external physical changes. Thus arose those striking characteristics and special modifications which still distinguish the chief races of mankind. The red, black, yellow, or blushing white skin; the straight, the curly, the woolly hair; the scanty or abundant beard; the straight or oblique eyes; the various forms of the pelvis, the cranium, and other parts of the skeleton.

But while these changes had been going on, his mental development had correspondingly advanced, and had now reached that condition in which it began powerfully to influence his whole existence, and would therefore, become subject to the irresistible action of "natural selection." This action would rapidly give the ascendancy to mind: speech would probably now be first developed, leading to a still further advance of the mental faculties, and from that moment man as regards his physical form would remain almost stationary. The art of making weapons, division of labour, anticipation of the future, restraint of the appetites, moral, social and sympathetic feelings, would now have a preponderating influence on his well being, and would therefore be that part of his nature on which "natural selection" would most powerfully act; and we should thus have explained that wonderful persistence of mere physical characteristics, which is the stumbling-block of those who advocate the unity of mankind.

We are now, therefore, enabled to harmonise the

conflicting views of anthropologists on this subject. Man may have been, indeed I believe must have been, once a homogeneous race; but it was at a period of which we have as yet discovered no remains, at a period so remote in his history, that he had not yet acquired that wonderfully developed brain, the organ of the mind, which now, even in his lowest examples, raises him far above the highest brutes;--at a period when he had the form but hardly the nature of man, when he neither possessed human speech, nor those sympathetic and moral feelings which in a greater or less degree everywhere now distinguish the race. Just in proportion as these truly human faculties became developed in him would his physical features become fixed and permanent, because the latter would be of less importance to his well being; he would be kept in harmony with the slowly changing universe around him, by an advance in mind, rather than by a change in body. If, therefore, we are of opinion that he was not really man till these higher faculties were developed, we may fairly assert that there were many originally distinct races of men; while, if we think that a being like us in form and structure, but with mental faculties scarcely raised above the brute, must still be considered to have been human, we are fully entitled to maintain the common origin of all mankind.

These considerations, it will be seen, enable us to place the origin of man at a much more remote geological epoch than has yet been thought possible. He may even have lived in the Eocene or Miocene period, when not a single mammal possessed the same form as any existing species. For, in the long series of ages during which the forms of these primeval mammals were being slowly specialised into those now inhabiting the earth, the power which acted to modify them would only affect the mental organisation of man. His brain alone would have increased in size and complexity and his cranium have undergone corresponding changes of form, while the whole structure of lower animals was being changed. This will enable us to understand how the fossil crania of Denise and Engis agree so closely with existing forms, although they undoubtedly existed in company with large mammalia now extinct. The Neanderthal skull may be a specimen of one of the lowest races then existing, just as the Australians are the lowest of our modern epoch. We have no reason to suppose that mind and brain and skull-modification, could go on quicker than that of the other parts of the organisation, and we must, therefore, look back very far in the past to find man in that early condition in which his mind was not sufficiently developed to remove his body from the modifying influence of external conditions, and the cumulative action of "natural selection." I believe, therefore, that there is no *à priori* reason against our finding the remains of man or his works, in the middle or later tertiary deposits. The absence of all such remains in the European beds of this age has little weight, because as we go further back in time, it is natural to suppose that man's distribution over the surface of the earth was less universal than at present. Besides, Europe was in a great measure submerged during the tertiary epoch, and though its scattered islands may have been uninhabited by man, it by no means follows that he did not at the same time exist in warm or tropical continents. If geologists can point out to us the most extensive land in the warmer regions of the earth, which has not been submerged since eocene or miocene times, it is there that we may expect to find some traces of the very early progenitors of man. It is there that we may trace back the gradually decreasing brain of former races, till we come to a time when the body also begins materially to differ. Then we shall have reached the starting point of the human family. Before that period, he had not mind enough to preserve his body from change, and would, therefore, have been subject to the same comparatively rapid modifications of form as the other mammals.

If the views I have here endeavoured to sustain have any foundation, they give us a new argument for placing man apart, as not only the head and culminating point of the grand series of organic nature, but as in some degree a new and distinct order of being. From those infinitely remote ages, when the first rudiments of organic life appeared upon the earth, every plant, and every animal has been subject to one great law of physical change. As the earth has gone through its grand cycles of geological, climatal and organic progress, every form of life has been subject to its irresistible action, and has been continually, but imperceptibly moulded into such new shapes as would preserve their harmony with the ever changing universe. No living thing could escape this law of its being; none could remain unchanged and live, amid the universal change around it.

At length, however, there came into existence a being in whom that subtle force we term *mind*, became of greater importance than his mere bodily structure. Though with a naked and unprotected body, *this* gave him clothing against the varying inclemencies of the seasons. Though unable to compete with the deer in swiftness, or with the wild bull in strength, *this* gave him weapons with which to capture or overcome both. Though less capable than most other animals of living on the herbs and the fruits that unaided nature supplies, this wonderful faculty taught him to govern and direct nature to his own benefit, and make her produce food for him when and where he pleased. From the moment when the first skin was used as a covering, when the first rude spear was formed to assist in the chase, the first seed sown or shoot planted, a grand revolution was effected in nature, a revolution which in all the previous ages of the earth's history had had no parallel, for a being had arisen who was no longer necessarily subject to change with the changing universe--a being who was in some degree superior to nature, inasmuch, as he knew how to control and regulate her action, and could keep himself in harmony with her, not by a change in body, but by an advance of mind.

Here, then, we see the true grandeur and dignity of man. On this view of his special attributes, we may admit that even those who claim for him a position as an order, a class, or a sub-kingdom by himself, have some reason on their side. He is, indeed, a being apart, since he is not influenced by the great laws which irresistibly modify all other organic beings. Nay more; this victory which he has gained for himself gives him a directing influence over other existences. Man has not only escaped "natural selection" himself, but he actually is able to take away some of that power from nature which, before his appearance, she universally exercised. We can anticipate the time when the earth will produce only cultivated plants and domestic animals; when man's selection shall have supplanted "natural selection"; and when the ocean will be the only domain in which that power can be exerted, which for countless cycles of ages ruled supreme over all the earth.

Briefly to recapitulate the argument;--in two distinct ways has man escaped the influence of those laws which have produced unceasing change in the animal world. By his superior intellect he is enabled to provide himself with clothing and weapons, and by cultivating the soil to obtain a constant supply of congenial food. This renders it unnecessary for his body, like those of the lower animals, to be modified in accordance with changing conditions--to gain a warmer natural covering, to acquire more powerful teeth or claws, or to become adapted to obtain and digest new kinds of food, as circumstances may require. By his superior sympathetic and moral feelings, he becomes fitted for the social state; he ceases to plunder the weak and helpless of his tribe; he shares the game which he has caught with less active or less fortunate hunters, or exchanges it for weapons which even the sick or the deformed can fashion; he saves the sick and wounded from death; and thus the power which leads to the rigid destruction of all animals who cannot in every respect help themselves, is prevented from acting on him.

This power is "natural selection"; and, as by no other means can it be shewn that individual variations can ever become accumulated and rendered permanent so as to form well-marked races, it follows that the differences we now behold in mankind must have been produced before he became possessed of a human intellect or human sympathies. This view also renders possible, or even requires, the existence of man at a comparatively remote geological epoch. For, during the long periods in which other animals have been undergoing modification in their whole structure to such an amount as to constitute distinct genera and families, man's *body* will have remained generically, or even specifically, the same, while his *head* and *brain* alone will have undergone modification equal to theirs. We can thus understand how it is that, judging from the head and brain, Professor Owen places man in a distinct sub-class of mammalia, while, as regards the rest of his body, there is the closest anatomical resemblance to that of the anthropoid apes, "every tooth, every bone, strictly homologous--which makes the determination of the difference between *Homo* and *Pithecus* the anatomist's difficulty." The present theory fully recognises and accounts for these facts; and we may perhaps claim as corroborative of its truth, that it neither requires us to depreciate the intellectual chasm which separates man from the apes, nor refuses full recognition of the striking resemblances to them which exist in other parts of its structure.

In concluding this brief sketch of a great subject, I would point out its bearing upon the future of the human race. If my conclusions are just, it must inevitably follow that the higher--the more intellectual and moral--must displace the lower and more degraded races; and the power of "natural selection", still acting on his mental organisation, must ever lead to the more perfect adaptation of man's higher faculties to the conditions of surrounding nature, and to the exigencies of the social state.[2] While his external form will probably ever remain unchanged, except in the development of that perfect beauty which results from a healthy and well organised body, refined and ennobled by the highest intellectual faculties and sympathetic emotions, his mental constitution may continue to advance and improve till the world is again inhabited by a single homogeneous race, no individual of which will be inferior to the noblest specimens of existing humanity. Each one will then work out his own happiness in relation to that of his fellows; perfect freedom of action will be maintained, since the well balanced moral faculties will never permit any one to transgress on the equal freedom of others; restrictive laws will not be wanted, for each man will be guided by the best of laws; a thorough appreciation of the rights, and a perfect sympathy with the feelings, of all about him; compulsory government will have died away as unnecessary (for every man will know how to govern himself), and will be replaced by voluntary associations for all beneficial public purposes; the passions and animal propensities will be restrained within those limits which most conduce to happiness; and mankind will have at length discovered that it was only required of them to develope the capacities of their higher nature, in order to convert this earth, which had so long been the theatre of their unbridled passions, and the scene of unimaginable misery, into as bright a paradise as ever haunted the dreams of seer or poet.[4]

Notes Appearing in the Original Work

[1]M. Guizot says: "If we regard the immediate influence of climate upon men, perhaps it is not so extensive as has been supposed. But the indirect influence of climate, that which, for example, results from the fact that, in a warm

country, men live in the open air, while in a cold country they shut themselves up in their houses; that in the one case they nourish themselves in one manner, in the other in another;--these are facts of great importance, facts which, by the simple difference of material life, act powerfully upon civilisation." (*Hist. of Civilisation in Europe.*)

[2]"It is probable that the present state and condition of New Zealand exhibit more nearly than any other the condition of Britain when the Romans entered it." (Turner, *Hist. of the Anglo-Saxons*, i, p. 69.) "When the Romans first became acquainted with Germany, the natives had advanced but a few steps beyond the savage state." (*Encyc. Brit.*, art. Germany.) [3]M. Guizot says: "For myself, I am convinced that there is a destiny of humanity, a transmission of the aggregate of civilisation." (*Civilisation in Europe.*) [4]The general idea and argument of this paper I believe to be new. It was, however, the perusal of Mr. Herbert Spencer's works, especially *Social Statics*, that suggested it to me, and at the same time furnished me with some of the applications. *[[Editor's Note: An account of discussion following the presentation of the paper was printed in the original source. This is the text of that discussion.]]*

The President proposed a vote of thanks to the author of the paper, and the meeting passed it unanimously.

The following discussion then took place.

Mr. Luke Burke said: No one will be surprised at my saying that the lecturer has made the very best of his case. That would be naturally expected, from what we know of Mr. Wallace's antecedents. I have only had the pleasure of hearing one paper from him, but that has given me very great interest and very great respect for his talents. If it had been possible to make a good case out of the theory which has been proposed, Mr. Wallace would have done it; but, unfortunately, the case appears to me to be altogether hopeless. I have three fundamental objections to urge against his theory, and I will confine myself to these; although, of course, there are many minor objections that would occur in regard to incidental remarks. I must, however, not forget to say that the theory by which he accounts for the permanency of human forms as contrasted with the inferior animals is exceedingly ingenious; but, unfortunately, it assumes that one part of the organism can gradually be modified without the requisite correlations in the others. It divorces our power of judging of the mind from the body; and I affirm that we have that power mentally, and not necessarily from the shape of the head. If we had sufficient intelligence, from any one part of the body, we ought to be able to infer everything else, internal and external. We cannot. The cypher is there, only we cannot read it. However, the first objection I have to urge against the theory of Mr. Darwin is, that it completely loses sight of the real point at issue--that it does not state the proposition correctly. The point at issue is, not whether these various external influences--food, climate, exercise, etc.--are capable of producing modifications; though, even there, I am perfectly ready to meet it. But the point at issue is this, Can they produce the modification actually required? Can they change one set of harmonious forms into an absolutely different set? Can they change one mechanism into another? Can they change that wonderful mechanism which you call wolf into that other equally wonderful and distinct set of proportions which you call greyhound, poodle, or spaniel? It is very well for Mr. Darwin to say that changes in one part of the frame will induce changes in the other. I agree with that, because it is done by organic laws; but you might as well say that a change in one part of a watch would superinduce the change in another. Yes, if the change is made by the watchmaker. That is quite another thing; and the question we have to determine is, What will change one kind of mechanism into another? In the body of the greyhound there is not a single particle that remains in the same relation as in the body of the wolf; and yet each one is an instance of the most admirable mechanism. That is one point at issue. Then, again, in causation there are two essential ideas--the fitness of the instrument, and adequate power to work it. Now, it is perfectly unphilosophical to assign causation where you are not able to show fitness, unless you are able to prove causation as a matter of fact by other means. No one has attempted to do that; no one can do it. No one can show that the accidental agencies of climate, food, etc., can produce correlated changes in any case whatever. That is not proved as a matter of fact, and you have no right to assume it. For instance, food, when conveyed into the stomach, is converted into blood, and sent as blood to all parts of the system. That is a general action; but can you see anything in food that will lengthen a man's leg, shorten his waist, or *vice versâ*, or that will give him a small head or a large one relative to his body? Is there anything in food or climate that can do that? Why, we have not yet been able to prove that climate changes the colour of races, except temporarily, by producing vesicles, etc. That is the second objection, therefore, that in this theory there is no conceivable fitness in the assigned cause to produce the assigned effect. Next, I maintain that it is absolutely impossible that these causes should produce such effects. The fundamental law of the universe is the law of causation. That law is, that there is an inevitable relation between the cause and the effect; that, as causes vary, so must effects vary. If, then, you want to know the unknown cause of a given effect, all you have to do is to find out the known cause of some analogous and similar effect, and then you know that there is a corresponding difference between the causes as between the effect, and also a corresponding resemblance. Now, then, here is the cause of mechanism. All mechanism is one in principle, whether living mechanism or the mechanism produced by man. All imply correlation of parts and functions--adaptation of means to ends. Now, then, do we know of any cause that is competent to produce such things? We do. Intelligence is competent; we see human intelligence doing such things. No cause in the universe except intelligence, then, can produce effects anything like those of intelligence. Surely non-intelligence cannot do it. Surely a non-intelligent cause cannot produce an intelligent effect. And not only so, but intelligence can never act without producing such things. Man never acts intelligently without adapting means to ends. Here, then, we have a case in which mechanism and all the wonders of mechanism are producible by a known cause; consequently all the mechanism of the universe is, argumentatively, the result of intelligence. If, then, we want to know how species originated, we must go forth to those parts of nature where everything is regulated by a determined plan. I will tell you of a case in which you may change types very easily--in a single generation; you do not want infinite time. The simple crossing of types. The crossing of races produces intermediate races, and they live and exist. Very well, there is a cause; but that is out of the bounds of the theory of natural selection. That has nothing to do with Mr. Darwin's infinitesimal working. Here comes the difficulty;

the crossing of races is rigidly barred within fixed limits. What are you to do out of those limits? How do you get types, then? By a mixture of different breeds of dogs you can get different types and varieties of dogs--some beautiful, some incongruous. Mr. Darwin's theory is admirable for telling us how races die out, but I do not see that it tells us how races come in. That is the point. Well, the crossing of dogs will produce--what? A cow? How does the cow come? Again, Mr. Darwin's theory requires us to start with the species before there can be anything like a change; but how did the species come? How did the first type come? Well, then, I say that the types outside the bounds of crossing come just as the first types come--by the plan of nature. There is one way of perfectly understanding it. In the living organism, you know that the various structures and portions have all their separate organs; you know that a muscle does not develope into a nerve, and that a nerve does not develope into a lung or into blood-vessels. Not only every muscle, but every nervous fibre has its own origin. Well, call this great globe--this cosmos in which we exist--call this an organism, and you have the whole affair. By the laws of that organism, by the plan inherent in that organism, the first type came. The next type came at its pre-determined moment, when a certain state of cosmic influences were provided; just as in the living organism bone never appears before a certain time, just as the brain does not appear before a certain time; or in the world's organism, as geology reveals to us, there are periods when there are only slight changes, and then all of a sudden we come upon entirely new types. You see it is no infinitesimal sliding. Yes, there are a number of contemporaneous forms that present a great number of shadowings, but that is co-existence. You have not shewn the sequence. This, then, is a point at issue. What is it that produces diversities beyond the bounds of species--germs, if you choose to call it so? What was it that originated the first species? I could very easily enter into the question of the varieties produced in the ordinary course of things, but they must all be within the race. They are not varieties beyond the bounds of species. The varieties that take place in the ordinary course of parentage only imply the growth of the species and type; for every type has its life, like the individual. The laws of life are always the same; and consequently types are born and are developed in the succession of generations as a matter of necessity, and then they die and pass away. These, then, are the points we have to examine in the theory. What produces mechanistic changes, and what produced the first type, and what produces the types outside the process of intermixture?

Mr. George Witt: I really have not understood the gentleman who has taken up so much of the time of the meeting. It reminds me very much of the Scotchman's definition of metaphysics: excuse me if I repeat it. "When the party who listens disna ken what the party who speaks means, and when the party who speaks disna ken what he means himsel--that is metaphysics." (*Laughter.*)

Mr. Burke: There is evidently one person who cannot understand, at all events. (*Laughter.*)

Mr. S. E. B. Bouverie-Pusey: We have listened to a very eloquent attack on the transmutation hypothesis in general; but I understood that Mr. Wallace did not mean so much to bring that doctrine forward, as to show that, assuming its truth, it would easily explain the phenomena of the races of man, their gradations into each other, and their present permanence. What we are told by Mr. Burke is principally that you can produce variations within the limits of species, but not outside; but that assumes the question whether there is a difference of kind, or species, or variety. Mr. Darwin does two things. He shows how varieties are produced--that is, by the action of natural selection; and he proves (at least in the opinion of many persons, myself included) that there are differences between species and varieties; and, as we know that varieties may be produced by natural selection, we may presume that in a sufficient length of time species and genera may be produced. Some say that it extends to the origin of the universe; but that does not follow. Many suppose the universe to be the creation of one Deity, some of opposite principles; but Darwin does not teach anything of the kind. The whole question raised by Mr. Burke is not touched by the Darwinian hypothesis at all. Mr. Burke has told us very fairly, that we ought to attribute things to such causes as we see in operation. Darwin and Mr. Wallace believe themselves to have proved that natural selection is such a cause. I must confess, however, that the idea in this paper was totally new to myself; and I believe that it must strike every one here as constituting a new era in anthropology.

Mr. T. Bendyshe: The eloquent discourse we have heard from Mr. Burke, has nearly driven out of my recollection the ingenious paper we previously heard from Mr. Wallace. There are still some points, however, which I am able to recollect, and on which I cannot altogether agree with the author. As far as I understood it, the principal scope of the paper was, that in proportion as the intellect of man became developed, he was enabled to triumph over every climatic influence. Now if one thing has been proved more than another about the race of man, it is this, that the inhabitants of temperate climates have been unable to live and flourish either in tropical climates, or in the polar- -the hyperborean climates; and *vice versâ*. If, therefore, all the intellect of the European is unable to give him the slightest footing whatever in the tropics, what becomes of

Mr. Wallace's proposition? This is not a question of natural selection on the struggle for existence between one animal and another of nearly allied species; this is a struggle of an animal with climate. I think that Mr. Darwin in his book has some expression of this kind. He applies the doctrine of Malthus with redoubled force to the animal kingdom. Now the doctrine of Malthus begins with the statement, that any animal or plant, if not checked by others, would in a short space of time cover the whole surface of the globe. He says that is incontrovertibly true. Now I should be inclined to say, that it is unquestionably false, that on the Darwinian theory, any animal could only cover the globe in process of time if uninterfered with, by ceasing to be the same animal or plant. That is the outside of what any one would admit from Darwin's theory. The very principle of that theory, Mr. Darwin does not exactly see the consequence of. It is not the theory of the struggle of existence between one animal and another, and, therefore, the idea that man, in proportion as his mind becomes developed, is able to overcome all climatic difficulties, is quite contrary to all observed facts. If it be said that the mind of the European is so extremely developed, that he has now lost the power of controlling his physical body--that the pendulum has swung so far that he cannot get it back, how is it that he can produce no effect upon those races of men who certainly have not been developed far beyond the animal, the negro, or the inhabitant of Tierra del Fuego? The intellect of the European applied in every possible manner to enable these beings to live outside the zone in which they are born, can no more make them flourish than his own progeny. He can produce no effect on them. They perish in a temperate, just as much as he perishes in a tropical zone. Then again, man in his progress to the highly intellectual European, supposing him to be the descendant of one original tribe or parent, has, we have every reason to believe, passed through all these phases; that he has passed through a tropical epoch, a glacial epoch, a temperate epoch. Now, how is it, if our predecessors have gone through all these forms, that we are incapable of existing in one of those climates in which our ancestors have actually lived? There again the theory fails, and I was unable to see anything in Wallace's paper that would answer this objection. In fact, in his paper, as in the book of Darwin, the struggle for existence has not been contemplated as applying not only to the contest between one animal, and a nearly allied animal, but to other species. It has been considered merely in that light, and not as a struggle, which any animal must have with climatic conditions, if it wishes to spread itself as Mr. Wallace seems to think, an entirely homogeneous race may do, over the whole surface of the globe.

Mr. Reddie: Having recently given my opinion as to the theory of the origin of species at some length, in a paper,

I am only anxious now to ask one or two questions of Mr. Wallace, because I should like to have this theory fully developed. But I may observe that I think he has raised a false issue in trying to connect the varieties of one species of living animals with Mr. Darwin's theory, which has nothing to do, strictly speaking, with varieties, but with the "origin of species"--not of varieties--by natural selection. I will not go into the speculative details which Mr. Wallace has very eloquently put before us as regards an imaginary world, which I think were extremely Utopian, and which, when this paper comes to be read and compared with all our experience of the history of the human race in historical periods, will, I think, be found totally inconsistent with all the facts of man's experience. For example, about the cold climates;--those who lived in the coldest climates were to have the best houses and clothes. Then compare the Esquimaux and the English--why, the thing is absurd. But I do not want to go into these details, because they lead us, I think, very wide of the main question. He told us a great deal about man--man, however, as far as I could make out, before he was man, because it was when he had no intellect or speech--and he expressly told us that the intellect of man and his speech became developed about the same time. Then what I want to know is, upon Mr. Wallace's principle, or any other principle of "natural selection", how this intellect came at all? We have the animal--something I suppose between the man and the gorilla--but it could not speak or think. From whence did this intellect, then, proceed at all? He gave us formerly something new in Darwin's theory, when he told us that the development of the canine teeth was not due to animal food, but to fighting for the females! But I think the Utopia of the past, was nothing compared to the Utopia of the future, as painted by Mr. Wallace. Mankind began a homogeneous race--he did not tell us whether a white or a black race--and it is to end a homogeneous race; and we are all to be so wise, that there are to be no wrongs or evils! Meantime, I shall be glad to hear Mr. Wallace explain how intellect was developed according to his theory in this curious being, whom I do not know how to describe, except by calling him "man before he was man".

Mr. Carter Blake: The most able paper of Mr. Wallace has given so clear an account of his theory, and Mr. Bendyshe and Mr. Pusey have so clearly expressed some of the criticisms I intended to have made on it, that I shall not detain you for a long period. One or two of the points to which Mr. Wallace called attention are, however, still open to debate. With respect to our knowledge of human history, is it a fact that the nations that have been extirpated by other nations, whose ethnic eras have been followed by other successive nations--is it a fact that they were inferior, either intellectually or physically, to the nations that came after them? Let us take an example in the case of the Basques. The Basques have been almost entirely extirpated from Western Europe. At one time, they occupied a large area; while at the present time, they are confined to very limited areas in Spain and France. But we know absolutely nothing about the history of the Basques, and we are not entitled to affirm that they were in any way inferior to the early savage Teutonic or Celtic nations that immediately extirpated them. This seems an important objection to some of the instances which Mr. Wallace has brought forward. Again, let us take the instance of the Celtic nations. We know that the Celtic nations, especially the Gauls, were driven westwards by the Frank or Teutonic nations; but if we compare the early traces of civilisation, which are afforded to us by the evidence of the most reliable contemporary historians, we know that the early Gauls, at least during the Roman period, were in a far higher degree of civilisation than those Franks who ultimately drove them before them, and who now occupy so large a portion of the French and Western German areas. There seems, in point of fact, to have been no intellectual inferiority between the Celtic and the Teutonic nations, and also no physical inferiority. It is true, that if we take some few striking examples of Scandinavian skeletons and measure their height, we see that the Scandinavian nations are those that usually comprise men of great stature, but when we take a fair average, not upon the whole higher or stronger men than those of the indigenous Celtic stock. There seems to have been no physical superiority of the Teutonic nations, and therefore when we apply this theory of the extermination of weak physical frames in the struggle for life--which struggle has undoubtedly operated in those inferior types of men (inferior as they were at that time) I fail to see what is the object that this theory of natural selections effects as to the extermination of these forms of life in Western Europe, so far as history gives us information on the subject. Then, with respect to there being a certain correlation between the structure of man and the locality in which he lives, if we examine a great many tribes of men at this time, there is not the slightest correlation between the structure of man and his habitat. For example, in the tropical countries we have certain races with a thick skull, and there have not been wanting theorists--I will not call them anthropologists--that have imagined that such thickness of skull was given to those nations as a beneficent provision to enable them the better to survive under the burning sun. Such is one version of the story, and I fear that the advocates of the theory of natural selection would adopt a similar style of argument. They would tell us that there are men of a certain average thickness of skull, in warm climates that those men who had a skull of greater thickness would in process of time survive, and that the thin-skulled races would in process of time die out. Well, such a thing may have some foundation in truth. But in

India, where the sun is as torrid as in any other part of the globe, we find a nation that has the thinnest skull. I confess, therefore, that I do not see the connection between the structure of superior animals and the circumstances in which they live, any more than I see in all cases the connection between the adaptation of the structure of the inferior animals and the circumstances in which they live. Anthropologists will in the course of time adopt this style of argument; and as to the reference which has been made to final causes, that, I think, is quite a bygone style of argument. Then Mr. Wallace has hinted that we may, perhaps, be entitled to consider man as the commencement of an entirely new order of things. This may be so. Of course, we cannot say when a new order of beings may commence or end. But what are the proved facts? That man is more like the inferior animals--at least, more so than anything else on earth; that, taking the arguments of the transmutationist on its lowest, most generalised, and most simple aspect, man is a great deal more like the gorilla and chimpanzee than the whale, or than any hypothetical sort of animal that may belong to a new order of beings. Then, with respect to man controlling nature. I do not know how he does so. It appears to me that he is subject to just the same diseases and vicissitudes of climate as inferior animals. The drought or the loss of food that exterminates the inferior animals exterminates man; and I do not see how man can be excluded from simple physiological laws, by saying that civilisation controls nature. Of course, it does to a certain extent; but civilisation has been utterly inadequate to take man out of the power of ordinary diseases, and those climatic effects which influence human beings as well as inferior animals. Having made this criticism, I hope that these observations will not be taken as against the theory of transmutation of man from the inferior animals. That theory has great probabilities in its favour, and will no doubt be borne out by facts. Whether the Darwinian theory can help us is another question; and, in the meantime, such papers as Mr. Wallace's will be in the highest degree valuable. I am sorry that his propositions should have been so remarkably misrepresented as they have been this evening. The whole theory of Mr. Darwin seems destined to pass through an age when it will be utterly misconceived and misrepresented by the general public, and a great evidence in its favour appears to be the amount of misrepresentation and divergences in the different versions, and that are placed in the scale respecting it. In respect to Mr. Burke's remarks, I shall not detain you very long. Mr. Burke commenced by saying he would lay down three general propositions. I did not understand what they were, but I mentally classed his remarks under three distinct heads--the statement of facts which I accept, of facts which I deny, and of facts which I did not understand. I will begin with facts which I accept. He has told us that an animal like a dog or a wolf never produces a cow. Mr. Burke and I are in perfect accord upon that topic, and I doubt not that Mr. Wallace and Mr. Darwin will be also. He also tells us that he never knew a nerve to develope into a muscle, or into lungs, or blood-vessels. Neither did I; and I believe those are the two principal facts of Mr. Burke, which I accept most unqualifiedly. But then he has told us what are the fundamental laws of the universe applied to man. I am sorry I don't know them, and I humbly doubt if any of us know them--we are here this evening as a society to try and discover some of the laws which regulate man. I for one, do not know what those fundamental laws may be, that may hereafter be discovered. Mr. Burke has also compared man to mechanism, and carried out the old illustration of man and the watch, showing that if you attack the mainspring certain consequences will follow. Gentlemen, the day is utterly past and gone when such an argument could have the slightest value in biology. We know that nothing that lives and moves and has its being in nature, bears the slightest analogy to mechanism in any way. Mr. Burke has told us that there are certain limits within which we can say that the hybrids are, or are not fertile in the human species. I for one, must deny this. I know not whether Mr. Burke, who knows the fundamental laws of the universe, may have some special information, but all the evidence which Broca and the best French authorities, or their brother anthropologist of America, Dr. Nott, can bring to bear, tells us distinctly that we cannot predict the limits in which hybrids are or are not fertile. The time will come, I doubt not, when we shall be able to do so; and a work will soon be laid before us, translated from a memoir by the secretary of our sister society in Paris, which will give some known facts on the subject. Till then, I submit, it is waste of time to discuss it.

Mr. Burke: I can only say, gentlemen, that I was bound in my address to give argument, but I was not bound to give understanding.

Mr. Pusey: I do not want to occupy the time of the Society, but it occurred to me that the fact of the congregation and yet non-transmutation of the human race, might possibly be explained by supposing, on Darwin's hypothesis, that he proceeded from one stock, but that he is now separated into different species. We do see species in the lower animals approaching one another--we see dogs, for instance, approaching to the wolf; but we do not see species ever transmuted into one another. But if we suppose distinct species to have had a common origin, the transmutation hypothesis might account for the facts.

The President: Before I call upon Mr. Wallace for his reply, I will make a few observations. I was, in common with yourselves, charmed with the paper; indeed, I was so much charmed, from the elaborate promises made in the opening of the paper of what "natural selection" could do, that a feeling of disappointment came over me at the conclusion, that those promises, which we were told would set to rights the difficulties of anthropologists, were not quite verified. When the author asserted that those difficulties would be set to rights by the principle of "natural selection," I do not think he sufficiently weighed the evidence that warranted him in making that assertion. I think it a pity that the two subjects of Darwin's hypothesis and Mr. Wallace's paper should have been so mixed up this evening; it would, perhaps, have been better if we had confined our remarks to subjects touched on in the paper. It appeared to me that the paper we have heard dealt very largely with assumptions. Mr. Wallace told us that man may have sprung from one race; indeed, he goes further, and says he must. Now, really this seems to me to be hardly a satisfactory argument. I hardly could have expected that the theory which was going to solve all the difficulties would at once make such an assertion, and I could not discover in the whole of the paper any facts that warranted the assertion. There is no doubt that hypotheses like Mr. Darwin's, and the one brought forward this evening, have a very great charm, because they attempt to explain so much. Does Mr. Wallace attempt to found his theory on known facts? If he does, then he failed to give those facts in his paper, and I am under a very strong impression that he has no facts to bring forward.

Mr. Wallace: What facts?

The President: Mr. Wallace asks me to specify the facts I allude to, and I have no objection to do so. Now, what do we learn from archæology? Take the whole of the remains of different continents, and what do we find? Go to America, and what do we find there? Do we find any indications of a different race dwelling there from the race of men that now exists? Not at all; and so wherever we go. Of course, if you go and take a Neanderthal skull as a type of a race, although there is good evidence to believe it simply the skull of an idiot, you beg the whole question. Mr. Wallace's theory appears to me not to be warranted by our present knowledge, and we cannot, I think, accept it. If the object of the paper is to assist in founding a science, that does not appear to have been carried out in the eloquent appeal which has been addressed to-night to the imagination. I must say that the opposite side has been equally imaginative. Mr. Burke, for instance, pronounces the thing to be impossible--a statement that is of course equally absurd. Assertions on either side stand for just nothing. And then the author of the paper tells us that man must have existed from a very remote period--the author says ten millions of years. Well,

we have, of course, no objection to that; any quantity of
time is at the disposal of any speculative philosopher. And
then he brings rather a charge against anthropologists--that
they look to that portion only of the truth that is on their
side, and insist on looking at the errors on the other side. I
hardly think that such a statement is fair to anthropologists,
ethnologists, and ethnographers; on the contrary, I believe
there are many anthropologists living who are at least
as capable of looking at the whole facts as any disciple of
Darwin. I think there are men in Europe who do not simply
look at facts which favour their own side, but who look at
facts as a whole, and look at them fairly, and endeavour to
interpret what may be truth from a careful examination of
the whole evidence. We are told that the Portuguese and
Spanish retain their characteristics in South America. That is
an assertion which ought to have some evidence to support
it. We are told that the Jews everywhere remain the same.
I think this is an argument that Mr. Wallace puts into the
mouth of a polygenist.

Mr. Wallace: Alike in features.

The President: If they are alike in features they will
be alike in other characteristics. This is no evidence
at all. I am perfectly aware that there is no change
in craniological development and stature, and the

mere change in the colour of the skin is temporary.

Mr. Bollaert: They lose their prolific character.

The President: Yes, on removal to climates that do not suit them; just as you cannot propagate a European race in India. Then he tells us that the best of the argument is for the principle of the diversity of the human race; and no doubt the polygenists will be glad to hear that they have the best of the argument. Now, Mr. Wallace very frankly admits, in opposition to some of the recent disciples of Mr. Darwin, that man differs from the ape very little in physical structure. I believe that some of his disciples now have come to say that there is a very great difference, and that a Neanderthal skull only approaches very little towards the ape. It is a pleasant thing to find one Darwinite, at least, true to his colours, and not frightened away from them by the clamour of the mob. Then he tells us that a hardy and more prolific race will be developed--a very provident race, too. I don't know, by the way, the physical characters of a provident race. I should be glad to know how this provident race is going to be produced? And then we have the statement that the Mexican government came from the north: but that is open to discussion, like all the other statements. Again, there is another assertion: that the ancient Britons were in a savage state at the time of Julius Caesar. Is that really a fact? Has it any but the barest traditional historic evidence as a foundation? It is not founded on known facts: but on tradition called history. It is brought forward as an argument to say that the Britons were slaves and savages two thousand years ago, and therefore that some people that are savages now will in that time be equal to us. But the whole thing is an absurdity, inasmuch as you cannot prove the fact, except on the barest traditional evidence. We were told of "natural selection" by virtue of external causes; now we are told of the inherent power; but this is surely wrong. There must be some mistake here, because the principle of selection is based on external circumstances. I should therefore expect Mr. Wallace, for the benefit of his argument, to withdraw the expression "inherent power." As to man being without the faculty of speech, I thought that speech was man's distinguishing characteristic. Professor Huxley, following Cuvier, at least says so. Then we are told that man can take away the power of natural selection. Well, if man can do that, what a powerless thing natural selection must be. If man, little man, even civilised man, has the power to take away this so-called law of natural selection, what a powerless law it must be. At the same time, I would say nothing against the law of natural selection as an hypothesis. It stands on its own merits as a purely philosophic speculation, but forms no part of inductive science. We ought always to make a great distinction in that. I put the Darwinian hypothesis just in the same category with any other hypothesis that can be brought against it on the same subject. Neither is more acceptable than the other, and it is only a question which can be proved. However, in all these matters we like a little poetical license; and I must confess that I listened with some pleasure to the beautiful dream that the author of the paper called up at the end. Although he did not satisfy me with science and with facts, he thoroughly satisfied me with the after-destiny of man. But the curious part of the case was that man's external characters were always to remain the same. That I do not like, and think that is a mistake. But his mind was to be advanced and improved without any development at all of the brain. All the other characters were to exist, though there was to be no individual inferior to the existing highest races. Well, that is satisfactory for some of the lower ones; they will not exist at all events. Then Mr. Wallace said that we were all to be equal; but that seems to be a long time off. Again, government will be unnecessary. Of course, that is a great blessing, I admit. Passions will not exist; or they will be ordered in a temperate manner, and exactly in accordance with man's physical formation. And all this is to be with exactly the same brain organisation as now. I suppose the laws of natural selection will entirely change the whole functions of the brain, and the whole functions of man will be changed, although his physical character will remain the same. Now, I hope that the author of the paper, for his own credit, will withdraw the whole of this dream, and not mix up these two subjects. As students of science we must object to this sort of dreaming, because it cannot be based on evidence. Some members of this society are accused of bringing forward speculations; but none of them have yet brought forward anything a thousandth part as speculative as this. I do hope that Mr. Wallace will make us understand that he does not insinuate that this dream has anything to do with his theory, or with Mr. Darwin's hypothesis, and then, I am sure, we shall all be very much indebted to him for coming before us this evening. Although I may regret that his own theory has not been better established; yet his paper shows most conclusively the exact position of the present state of Darwinianism. I believe this is the first occasion in which we have had a clear logical statement of the position in which the theory of transmutation by external circumstances now stands in reference to Anthropology; and I am sure you will all agree with me in heartily thanking the author of the paper.

Mr. Wallace: Before I begin *seriatim* to notice a few of the objections made to my paper, I should like to correct a slight misapprehension which Dr. Hunt has made, while fresh in my memory. I have been obliged, in order to compress my remarks, and at the same time to make my meaning clear, to use expressions which are, perhaps, not logically accurate. In the latter part of the paper, the argument is the contrast between change of body and change of mind. By the former was meant change of organisation, of the limbs particularly, and of other external physical characteristics. By the mind I always include the brain and skull--the organ of the mind--the cranium and the face; and therefore, when I afterwards contrasted change of external form with change of mind, of course I do not mean to say that the cranium which contains the organ of mind was stationary. Therefore, I beg to be understood that there is no contradiction in my argument,--that man may advance to this high state of civilisation, while his physical frame remains unchanged. Mr. Burke's observations have, to a great extent, been answered by several speakers. I would say that they appeared to me totally to misrepresent the purport of my paper. Of course it was seen that, to a certain extent, it was impossible to go into details with respect to this subject of natural selection, and I only brought forward my illustrations of it to refresh the memory of those who are not thoroughly acquainted with the whole theory. I do not now argue generally for that theory; I merely show how it applies to a particular doctrine of anthropology. I endeavoured to apply it in a way in which it has not been applied before. I will now pass on to notice the special objections that have been brought forward to my theory. Mr. Burke's arguments were all against the theory of natural selection itself; but Mr. Darwin has argued it so well that it is impossible for me to add anything. The two next gentlemen who spoke agreed with me generally. Mr. Bendyshe objected to my statement, that man, to some extent, triumphs over nature; and he argued that man does not triumph over climate, because Europeans cannot live in the tropics, and the natives of the tropics cannot live in Europe. First, I say that there are facts to show that this is not absolutely the case. There are cases in which Europeans have gone and resided in the tropics, and, as far as we can see, live there to this day perfectly well. One particular case I will mention. In the interior of South America, on the eastern slope of the Andes, the head waters of the Amazon, there is a district quite isolated from the rest of the world, cut off on the one side from the Pacific by the Andes, and on the other side by the intervention of Brazil, no communication of any kind having been allowed till recently. In this upper valley of the Amazon there is a large population, purely European, or at least very nearly so. There are a number of towns and cities there, numbering ten, fifteen, and even twenty thousand inhabitants. No doubt the race is partly mixed,--we cannot say how much, but my friend, Mr. Spruce the botanist, describes them to me as actually whiter than the Brazilians, remarkably white for a south European race. He was astonished to come upon so large a population, which knew nothing of any other part of the world. They are the descendants of some of the Spanish settlers. Here, then, we have the case of a European population transferred to a tropical country.

Mr. Bollaert: Will you name some of those cities?

Mr. Wallace: Well, Tarapoto, and Moyobamba.

Mr. Bollaert: I should say that the population was two-thirds Indian; certainly a mixed race.

Mr. Wallace: I got the information from a gentleman who has resided there, and he assured me that the mass of the population was white.

Mr. Bollaert: There is a great deal of Indian blood.

Mr. Bendyshe: What is the altitude?

Mr. Wallace: Not more than a thousand feet above the sea. The plain of the Andes is perfectly flat.

Mr. Bollaert: If there is Indian blood there, that is the very point.

Mr. Wallace: But it is urged, that directly you get a cross you get infertility; and yet here there are immense numbers.

Mr. Bollaert: I doubt extremely the immense numbers.

Mr. Wallace: I can only give the facts as they were given to me. If they are wrong, they can be disproved; but the question does not depend upon that; for admitting that man may not be able to stand a sudden change in climate, yet, supposing that the change were a slow one--supposing that Europe were gradually sunk beneath the sea from the north, so that we were gradually shoved, as it were, into a tropical climate at the rate of a few miles in a century,--do you not think that natural selection would act so that the race would stand the climate? I do not think we should all die out. All the facts of nature seem to be opposed to such a supposition. The dog has stood all over the world with us notwithstanding the climate.

Mr. Blake: May I ask the historical evidence of the migration of dogs?

Mr. Wallace: I cannot now go into that. Dogs are carried by man all over the world.

Mr. Bollaert: And they die.

Mr. Wallace: Mr. Reddie began by saying that Mr. Darwin's theory had nothing to do with varieties. Now, from my study of the theory, it appeared to be all founded on the study of varieties. The whole argument is based on varieties, showing that they merge gradually into species.

Mr. Reddie: What I meant to say was, that it was not limited to varieties.

Mr. Wallace: I thought you said it had nothing to do with varieties. Then, another strong argument was that the Esquimaux, notwithstanding their bad climate, do not build good houses, not so good as Englishmen. I have asserted that man, in his progress from a low to a high state, would be assisted by the necessary discipline of a harsh climate, which would make him exert his mental faculties much more than a tropical climate. Now, I think that is almost self-evident, and is not at all affected by the fact that the Esquimaux are less intelligent than the English. The question is, "Do they build houses at all?" Yes; and very good ones. Travellers describe how ingeniously they build their snow houses; and the manner in which they make their clothing and sledges shows that they are not so low intellectually as most of the inhabitants of tropical countries. Mr. Reddie also wants to know how the intellect came at first. I don't pretend to answer that question, because we must go so long back. If Mr. Reddie denies that any animal has intellect, it is a difficult question to answer; but if animals have intellect in different proportions, and if the human infant, the moment it is born, has not so much intellect as an animal, and if, as the infant grows, the intellect grows with it, I do not see the immense difficulty if you grant the universal process of selection from lower to higher animals. If you throw aside altogether, this process of selection, you need not make the objection about the intellect. Mr. Blake made a few objections, which may have some little weight. The principal was that we have no evidence to show that when one race, or nation, or people are exterminated, or driven out by another, the one that is so exterminated is necessarily inferior; and he wanted to show either by historic evidence or by remains of bodies that it is impossible to say that the Celtic was inferior to the Teutonic, or the Basque inferior to the race which drove them out. Now, it appears to me that the mere fact of one race supplanting another proves their superiority. It is not a question of intellect only, nor of bodily strength only. We cannot tell what causes may produce it. A hundred peculiarities, that we can hardly appreciate, may cause the one race to melt away, as it were, before the other. But still there is the plain fact that two races came into contact, and that one drives out the other. This is a proof that the one race is better fitted to live upon the world than the other. Mr. Blake says that there is no necessary correlation between man and his habitat; and he endeavoured to show that, by proving that the thickness of the crania does not vary in accordance with the heat of the sun. No doubt such an objection is very easy to make; but we must consider, is it at all likely that we shall be able, by our examination, to appreciate this correlation, whatever it may be. For instance, you take two animals; one lives in a northern hemisphere, the other in a southern,--one in a wet country, the other in a dry one. Can you tell me why these two animals are fitted to live in their respective climates? They may be so closely allied that you can hardly find out their differences; and if you cannot find out the difference in animals which serves to adapt them to the climate, is it likely you can find out the difference in man? But there are facts which show that there is a correlation between man and his habitat. For instance, take the case of the inhabitants of West Africa, who stand the fever and malaria of that country; and it is the same in

New Orleans. It is asserted in America, I believe, that one-fourth of black blood is enough to save the individual from the yellow fever in New Orleans. This is a striking case, I think, of correlation between man and his habitat. Then again, as to the prevalence of black-skinned races in the tropical regions, I do not believe that there is any special production of the black skin by the heat of the sun; but I believe that because the black skin is correlative to the hot sun, the black skinned constitution is best adapted to stand the diseases of the climate, and the process of natural selection has preserved them. If we find a people who are apparently not well adapted to stand the climate, we have some reason to believe that they are a comparatively recent immigration into the country. My friend, Mr. Bates, who is not here, has supported this theory from his observations on the Amazon, asserting that the inhabitants of tropical America are a recent introduction. He comes to that conclusion from a great many peculiarities of manners and customs, and if so, it is a corroboration of the argument that races do become correlated to the climate in which they live. Mr. Blake objected to my statement, that man can to a certain extent control nature. He asserted that man could not control disease; but that was not the point I went upon. I especially mentioned the point on which man can control nature,--raising himself by his intellect above the action of natural selection, which changes the forms of other animals, because they could only be kept in harmony with the universe except by being changed; whereas man is kept in harmony by his mind. Again, no weak animal--no animal born with a sickly constitution--lives to propagate its kind: but man does. Hundreds of weak individuals live to a comparatively healthy and comfortable old age, and have large families. This is a special case, in which man controls nature differently to the animals. He controls nature so much that he is an exception to all the rest of animated beings. Dr. Hunt made a great many special objections. He says, I disappointed him, because I promised to explain everything. I must say, I did not. I simply proposed to myself to explain, or rather to suggest, a theory which should do away with this difficulty of the absolute contradiction between two classes of ethnologists, commonly called the monogenists and the polygenists, by showing that both were right. I think that is a most satisfactory way of harmonising people that differ. Again, he objects to my using the expression "must have been". Well, I put in the words "I believe," and "according to the Darwinian theory", because, according to that theory, every group of species arises from one, every group of varieties from one, every group of individuals from a pair; therefore, if you do but go far back enough, you must come to a unity of origin. If that theory is utterly wrong, then my argument goes for nothing. Then Dr. Hunt says I did not give facts enough. Well, first you are aware that in a subject like this, if a sufficiency of facts were given, they would fill a volume; consequently, I was obliged in this paper to sketch and allude hastily to facts. Dr. Hunt asserts, that archæology shows that the ancient races were the same as modern. Well, that is a fact I quoted on my own side, and his quoting it against me only shows that you can twist a fact as you like. I quoted it as a proof that you must go to an enormous distance of time to bridge over the difference between the crania of the lower animals and of man. I said, perhaps a million, or even ten millions, of years were necessary. If my argument is correct, it is a logical conclusion. Dr. Hunt objects to my using an expression to the effect that students are rather dogmatic in assertions of this kind. Well, I think I could bring forward facts to prove this; and I should think that anybody who knows anything of the literature of the subject, would agree with me that there is the strongest feeling on both sides that they are right, and that they express their feelings in the strongest manner, and that each party is inclined to look down on what it believes to be the absurd ideas of the other. Still, I do not deny that there are some who do not manifest this dogmatic feeling. With respect to the fact about the Portuguese and Spaniards in South America, I can assert it on my own authority, because I have lived among them, and have seen European families in tropical countries who have been there for many generations. I may name the town of Amboyna, in the Moluccas, where there are families that have kept their blood pure for three hundred years, as fair skinned, and in every respect like Dutch men and women.

Mr. Bollaert: Have not fresh families been sent out to them from Holland?

Mr. Wallace: Possibly so.

Mr. Bollaert: But that is very important.

Mr. Wallace: I allow it; but still there is the fact, that this period of time has produced no change. If there was a change, notwithstanding a little fresh blood, it would be perceptible.

Mr. Bollaert: More than a little, depend upon it.

Mr. Wallace: But there is no perceptible difference. Of course, these kind of facts are the most difficult in the world to get at. You cannot isolate men. They will mix; and there is no possible fact you can bring forward but is liable to the same objection. It was thought at one time, by Prichard and the older ethnologists, that it was a strong argument for the unity of the race that the Jews were white, black, and brown. Now, it is known that in every case in which the Jews have changed colour apparently, it has been the Jewish converts who have been treated as Jews, simply because they have embraced that religion. But a better proof than colour is physiognomy, which you see maintained in the Jews all over the world. Physiognomy maintains itself much longer than colour; and it seems as if the physiognomy of the superior race maintained itself much longer than the inferior; whereas the colour of the inferior race is often most lasting. For example, I may mention the descendants of the Portuguese in the Malay archipelago. In a great many towns there are thousands of Portuguese; some of them keep the Portuguese language; others have lost it; but still Portuguese words crop up all over the land, and there are Portuguese customs and manners and European features; but still they are generally the same colour as the people of the country in which they live. With respect to Europeans not living in India, that is nothing when we remember what a vile climate it is. We live in it as an exceptional race; and if we could bring instances of the third generation, you would say there was mixed blood in them. Then again, Dr. Hunt wanted me to explain how I could use such a word as "provident". Why, is it not perfectly clear that if people live in a country where there is a severe winter, in which little or no food is to be had, that they must provide against the scarcity, and that gradually the race would become a provident race? Therefore, I think I am justified in saying that, given two races of the same capacity, and put one in a tropical and the other in a temperate climate, the one in the temperate climate will become the more provident race of the two. With respect to Britons ever having been savages, I cannot assert that; but I think it would puzzle Dr. Hunt to show that they were civilised. All the evidence we have proves that they were savages, as much so as the South Sea islanders.

The President: Chariots?

Mr. Wallace: The South Sea islanders had no horses. Well, then, as to the term "inherent," I do not mean to withdraw it. I mean to maintain it as a very proper expression; and the answer I gave to that last question about a provident race, will almost answer for this,--that peculiarities produced gradually by natural selection, or any other cause, become inherent. The very fact of the race being gradually brought into harmony with the climate of the country in which it is, gives it a superior power, and an inherent capacity to maintain it. I do not know whether the words are the same, but the sense is exactly the same as will be found in Darwin's own book, where he points out this extraordinary fact, the bearing of which had never been noticed before, that in Australia, in the Cape of Good Hope, and to a considerable degree in North America--in fact, to a great extent in all the comparatively limited areas to which Europeans go-- the weeds of Europe that are carried accidentally thrive and flourish there. They spread over the country, and maintain themselves in competition with the native weeds, showing that they are better adapted for the country than the plants which were apparently specially created for the country. Mr. Darwin explains it on his theory in this manner,--that Europe and Asia, which *[[are]]* now to a great extent dry land, have been long in existence as dry land; and that in the immense series of ages during which the changes of the northern continent have been going on, becoming modified from one form to another, sometimes to an inland climate, sometimes to a continental climate, sometimes a mountainous region, sometimes a flat region; owing to that great amount of change, its plants have acquired an immense variety of specialities: because, when a speciality is once acquired, it is not lost. It is handed down and kept in store, as it were, so that the immense mutations which the northern hemisphere has undergone, have given these plants a capacity of adapting themselves to a great variety of conditions. The result is, that directly they are carried into Australia these properties come into play. They have been adapted, in some previous state of the northern hemisphere, to similar conditions, and they have inherited this peculiarity by transmission, and therefore they are capable of driving out the plants of Australia merely by the inherent vigour they have gained. I applied this in illustration of the way in which civilised man has been developed by a great variety of circumstances. The intermixture of races has been very great. We are a mixed race to a very great extent, and therefore we have the capacities and powers of a great many; therefore, when we come into contact with the lower races, we are enabled in the same manner to drive them out. Then, it is said, that man without speech is not man. That is one of my points. I said, if you choose to consider he is not man, then so and so follows; but if you consider he is man, then so and so. And as to the argument, that if man could take the effects of natural selection away, it must be powerless,--that has not much to do with the subject. We might as well say, how powerless life is, because we can take it away,--when such a slight thing as stopping the mouth with pitch-plaster can destroy it. This only shows how easily it can be changed or destroyed; it does not prove its weakness. And so it does not show the weakness of natural selection, because man is able to modify it by putting himself into certain conditions, instead of leaving nature to select those conditions for him. I think I have now answered all the objections; and it is now so late that I really cannot detain you any longer. With regard to the poetical conclusion, I would merely say that I began it by stating that I would point out what I considered to be the bearings of this theory, if it is true. If it is not true, of course my remarks go for nothing; but I do not think myself that the concluding part of the paper is more poetical than true.

The meeting then adjourned.

www.ingramcontent.com/pod-product-compliance
Lightning Source LLC
Chambersburg PA
CBHW030135260626
47156CB00008B/2953